21 June 2006

To Ben, With the hope
what you will bend
your creative mind
to many other projects
in future!
With very best wishes
for your future.

David Wells

mindbenders and brainteasers

100 maddening mindbenders and curious conundrums, old and new

ROB EASTAWAY
AND DAVID WELLS

ROBSON BOOKS

This edition published in 2005 by Robson Books, The Chrysalis Building, Bramley Road, London W10 6SP

An imprint of **Chrysalis** Books Group plc

British Library Cataloguing in Publication Data
A catalogue record for this title is available from the British Library.

ISBN 1 86105 562 5

Typeset by SX Composing DTP, Rayleigh, Essex
Printed by Creative Print & Design (Wales), Ebbw Vale

INTRODUCTION

What's the point of a puzzle? For some, it is the enjoyment of working at a challenge and coming up with an answer. For others, looking up the solutions is at least as much fun as attempting the puzzles themselves. This is because the best puzzles often have surprising short cuts or unexpected twists in their solutions. In that respect, the solution to a good puzzle is like the punchline to a good joke.

This collection of puzzles includes many of our favourites. We've deliberately included a few that are old favourites, because we know that many people like to be reminded of the classics. The majority of the puzzles, however, will be unfamiliar to most readers.

Many of the puzzles can be solved after only a few moments thought – though sometimes the ones that seem simplest are the ones with the biggest traps. A few will require some serious thinking and some paper and pencil work, and these have been marked with a star.

We'd like to thank Jeremy Robson and Charlotte Howard for encouraging us to put together this new compilation, based on the books *Brainteasers* by David Wells and *Mindbenders* by Rob Eastaway and David Wells, originally published by Guinness.

* Puzzles with an asterisk are the ones most likely
to require some serious pen and paper work.

1

THE AARDVARK AND
THE ZEBRA

The *Illustrated Encyclopaedia* comes in two volumes, each of them about two inches thick. The books are sitting on the shelf next to each other, with A–M on the left and N–Z on the right. Josh has bookmarked the entries for *aardvark* and *zebra*. Roughly how far apart are the two bookmarks?

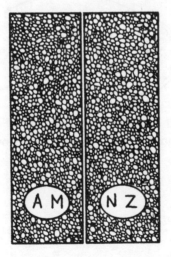

2

ROLLING WITH MONEY

Sue was fiddling with two 10p coins, rolling one of them round the other. Martin came over. 'You see these two identical coins?' said Sue. 'I am going to keep the one on the right still while I roll the one on the left round it without it slipping or sliding. What do you think will happen to the Queen's head when it gets to the other side?'

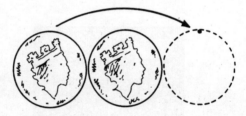

'The head will be upside down, of course!' Martin replied. Sue smiled smugly. What is the correct answer?

3

A WORD QUIZ

'I've got a strange list of words here,' said Sam, 'and apparently there is one missing.'

'Let me have a look at the list,' said Nicky.

CWM
FJORD
BANKS
VEXT
GLYPH

'A cwm and a fjord are both something to do with valleys, aren't they? And I suppose valleys have banks. What about glyph and vext though?' Nicky asked.

'Apparently a glyph is a mark like a hieroglyph, and vext is an ancient spelling of *vexed*.'

'Well I'm none the wiser,' said Nicky.

'Nor me,' said Sam.

Can you solve this quiz by finding the missing word?

4

TAKING THE BISCUIT

Tom had been meddling with his mother's biscuit tins while she was out. One of the tins contained only *Digestive* biscuits, another contained only *Rich Tea* biscuits, and the third contained a mixture. He had carefully taken off all the labels and replaced them so that each label was on the wrong tin.

When his mother returned, Tom confessed what he had done. She reached into one tin and removed a single biscuit. From that she could deduce exactly where all the labels ought to go. Which tin did she open?

5

PAST THE POSTMASTER

Mr Jones wanted to send a snooker cue by parcel post, but he was disappointed. The postmaster told him that it was a little over two metres in length, and that no parcel over 1.5 metres in length, width or depth could be accepted. Mr Jones was not stumped for long. He went away and wrapped it carefully, brought it back, and the postmaster accepted it for posting. He didn't have to shorten the cue. Indeed, it was just as long as before he had wrapped it. How did he get it past the postmaster?

6

A MOVING PROBLEM

This picture represents a cherry and a glass. Of course, you could put the cherry in the glass by picking it up and moving it, but that would be too easy. Your challenge is to put the cherry in the glass by moving only two matches.

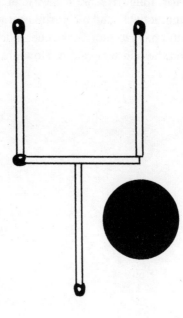

7

CHRISTMAS CARD MYSTERY

Last December the Crown & Greyhound pub was selling its own Christmas cards. Cards were sold separately and in theory you could ask for whatever number of cards you wanted. Among other combinations, many customers bought multiples of five cards: lots bought 5; some bought 15; and several bought 20. What seems odd, though, is that nobody at all bought 10 cards. Apart from coincidence, can you think of a simple explanation for this?

8

A KNOT OR NOT A KNOT?

Martin had found a length of old rope. 'Grab the other end and pull,' he said to Sue.

'Don't be silly,' she replied. 'It may turn into a knot.'

'No it won't,' said Martin.

'How do you know?' asked Sue.

'Bet you!' said Martin.

Was Martin right? Does the tangle of rope knot if you pull the ends?

9

TYPEWRITER TROUBLE*

Dear Mum,

Thanks for sending me your old electronic typewriter. It's great except for a strange little problem with the daisy wheel. Every time I type the letter z ju tubsut up hp gvooz. K owuv igv kv hkzhg zlir M gsqi lsqi.

Psxw sj pszj,

Xzk

Who sent the letter?

10

BUS QUIZ

On the town buses, all of the fares are either £1 or 70p. Yesterday, Mrs P put £1 on the tray and the driver gave her a £1 ticket. Then Mrs Q put £1 on the tray, and the driver asked her 'do you want a 70p or £1 ticket?'. The driver had never seen either of the women before, so why the different treatment?

11

WATER TEASER

The family were picnicking not far from the river bank at the spot marked X. Maria had gone to fetch some water in a can, but had forgotten her task and wandered off to pick some wild flowers at Y.

'Where's the water?' cried out Mrs Watson. 'I can't make tea without water!' Maria immediately put down her flowers and ran to the river bank, filled the can with water, and ran back to her mother. If Maria took the shortest route from Y to the river bank, then back from the river bank to X, where on the river bank did she fill her can?

Y
•

X
•

12

FUN RUN SPRINT

This is Angela's first time in the three mile Fun Run and she is finding it a little tough. In fact, she has just completed the first two miles, and as she looks at her watch she says to herself (breathlessly): 'I'm only averaging four miles an hour. Oh dear, I wanted to average six miles per hour for this run. I must go a bit faster.'

How fast will she have to run the final mile in order to get her average speed for the whole Fun Run up to 6 mph?

13

THE BEAR-FACED TRUTH

'Been on your trip?' asked Tuffer.

'Mm, yes,' replied Wimpish. 'Spent a week in a delightful spot, miles from anywhere, as peaceful as could be. And would you believe it, whichever direction I looked in, I was looking north!'

'That reminds me,' said Tuffer, 'of my last journey. On the very last morning I got up early, tramped ten miles south, turned and walked ten miles east, and then walked ten miles directly north, and I ended up exactly where I started. And would you believe it, standing in front of me was the biggest bear you have ever seen, so I shot it.'

Where did Wimpish go on his 'trip', and what colour was the bear that Tuffer shot?

14

MAGIC BEANS

In one of his notebooks, the famous artist and scientist Leonardo da Vinci describes the following trick. (To do the trick, you need a supply of small objects such as beans.)

'Take equal numbers of beans in each hand. Transfer four from right to left. Count the remaining beans in your right hand and cast them away. Cast away the same number from the left. Pick up five more beans. You will now have thirteen.'

In fact it doesn't matter how many beans you started with, you will always end up with thirteen in your hands. Why?

15

THE SECRET OF THE PENTAGON

How is it possible to make a regular pentagon, as shown, by tying a knot in a long strip of paper?

16

THE WOLF, THE GOAT AND THE CABBAGES

A man wants to take a wolf, a goat and a sack of cabbages across a river in a small boat, but he cannot take them all at once because the boat is only big enough to take him and either the wolf, the goat or the cabbages. Also, he can't leave the goat and cabbages alone, because the goat will eat the cabbages, and of course he couldn't leave the wolf and goat together because the wolf might get hungry. How did the man get all three across the river, without losing the cabbages or the goat?

17

FAIRGROUND CIRCLES

Janet started on the Big Wheel at the same moment that Barry started on the Twirler and Sheila started on the Rotator. The Big Wheel went round once every ten seconds, the Twirler once every seven seconds and the Rotator once every eight seconds.

If all three children started at the bottom of their respective rides, how long would it have been before they were in line with each other again at the bottom of each ride?

18

A REMARKABLE RELATIVE

'I saw my father's daughter's daughter's only cousin's father this afternoon,' remarked Gary. 'Delightful fellow, very intelligent, and jolly good-looking too, I must say.'

Who was this remarkable relative, and where did Gary see him?

19

SCARED CROWS

The crows were making hay in Farmer Giles's barley fields, so he got his gun and shot at them. He missed of course, but half of them did fly away, although ten of them soon came back.

Later that day, he went out again and shot at them. Half of them flew away and ten soon returned.

That evening he did the same and half flew away and ten returned. When he counted the crows, he discovered – to his disgust – that there was just the same number as there had been in the morning.

How many was that?

20

POCKET MONEY POSER*

'I will agree to increase your pocket money,' said Mr Street to his son, Dougal, 'but I will give you a choice. If you can tell me which choice is more favourable to you, then I will give you the increase that you chose. But if you cannot say which is better for you, I shall not increase your pocket money at all. Do you understand?'

'Yes, Dad,' said Dougal, who was tremendously polite and well-behaved.

Mr Street continued: 'I will increase your weekly pocket money by one third, and then give you an extra pound per week on top. Or, I will increase it by one quarter, and then add an extra £1.50 each week. Which would you prefer?'

Dougal thought for a while and then announced that it made no difference, whichever increase he chose, he would end up with the same amount of pocket money. 'Correct,' replied his father. 'So I shall give you an extra £3 per week.'

What was Dougal's weekly pocket money before his raise?

THE Z-SHAPED CUT

The carpenter needed to fill a hole in the floorboards. It was 2 feet wide and 9 feet long. Unfortunately, he only had one board, which was 3 feet wide and 6 feet long. It had the right area, but it was the wrong shape.

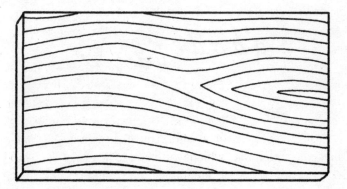

How could he cut the board into two pieces, which would then fit together to make a plank 2 x 9 feet?

22

ALPHABETICAL NUMBER

Which is the only whole number which, when written out as a word, has its letters in alphabetical order?

a b c d e f g . . .

23

JANE 4, MARY 1

Brian had two girlfriends whom he loved equally, so naturally he wanted to visit each one as often as he visited the other. To avoid making difficult decisions, he would just turn up at random at the station and wait for whichever train came first. Jane lived up the line and Mary lived down the line, and he knew that the up trains and the down trains both turned up at ten minute intervals.

Unfortunately, his plan was a failure. He found himself visiting Jane four times as often as he visited Mary, who was most upset and decided to ditch him.

Why didn't his cunning plan work?

MIRROR CODE

THIS PARAGRAPH CONTAINS A SECRET. THERE IS A SPECIAL WORD. IF YOU TURN THE PARAGRAPH UPSIDE DOWN AND LOOK AT IT IN A MIRROR, THE SPECIAL WORD WILL MIRACULOUSLY BECOME, AS IT WERE, DECODED WHILE EVERY SINGLE OTHER WORD, INCLUDING 'ZYGOL', WILL HAVE BEEN SLIGHTLY MESSED UP. CAN YOU FIND THE SPECIAL WORD (WITHOUT USING A MIRROR)?

25

GOING UP

Debbie lived on the sixteenth floor of a tower block. Normally she travelled with Janet, but one day she went out alone. She took the lift to the ground floor and caught the bus. On her return, however, she got the lift to the fifth floor only, and then walked up to the sixteenth floor. There was nothing wrong with the lift, and she would really rather not have had to make that long walk. What was the explanation for her odd behaviour?

26

LONG DISTANCE TRAVEL

Harry Hoppit, the famous solo airman, was in his London penthouse suite planning his next flight in his seaplane. 'Maybe I'll go over the North Pole,' he said out loud, looking out of the window towards the hills of Hampstead.

'On the other hand, I could go via America,' he muttered to himself, looking out towards Westminster Abbey and the Thames Valley beyond.

'Why don't you just take the quickest route?' asked his partner, Persephone.

'It makes no difference,' he replied. 'What's a few miles here or there, when I'm going so far?' and he continued muttering to himself, 'Perhaps I should fly over India . . .'

What was Hoppit's destination?

27

AN UNUSUAL BET

Janet had only £5 on her and she wanted £10 to go to the local cinema. She said to her brother, 'I will bet you five pounds that if you give me ten pounds, I will give you fifteen pounds.'

Her brother thought for a moment and accepted her bet.

Was he wise to do so?

28

OTHER SIDE, DEER

Charlotte found some toothpicks, arranged as the figure here shows.

Next to it were some instructions: 'This is a deer. You are looking at its right side. Move one toothpick so that you are looking at the deer's *left* side. It should be a mirror image of the picture you now see.'

Charlotte could see how to do it by moving two toothpicks, but not with one. She thought it must be a trick question, but when she did find a way to do it, she was surprised that it had taken her so long. Can you find the answer?

29

OVERTAKING TRAINS*

This picture shows the unfortunate situation after two trains found themselves on the same line, going in opposite directions. The problem the two drivers faced was how to pass each other, with no aid but the short spur line, which was only long enough to take either one truck or one engine at a time.

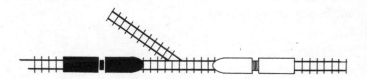

There was only one driving unit on each train, but there was nothing to stop part of one train being temporarily hooked to the other train and pulled by it or pushed by it.

How would you advise the drivers?

30

TAKING A RISE

Uncle Paul was watching a liner, which had just berthed in the harbour, with his niece Sally. 'Look at that ladder hanging down the side of the ship,' he said. 'The water is already only five rungs from the top. By high tide, it will be up to the top!'

Sally laughed loudly. How did she know at once that her uncle was having her on?

31

FAIR PRIZES *

Four girls at the back of the classroom were comparing the number of prizes they had won at the fair. 'I've got one more than you,' said Bernice. 'I've got two more than you,' said one girl to another. 'I've got three more than you,' said one to another, 'I've got four more than you,' 'I've got five more than you,' 'I've got six more than you,' rang out their excited voices, but we don't know who was talking to whom.

If they won a total of 27 prizes, how many did Bernice win?

32

A BORDER-LINE CASE

'How long is the boundary between Spain and Portugal?' asked Pamela, innocently.

'According to the Spanish map we used last holiday, it's 987 km,' replied her father. 'But the Portuguese customs man disagreed,' said her mother. 'He said it was 1,214 km long.'

'I think they're both wrong,' said Pamela. 'I think that with my little ruler' – at this point she waved her school ruler in the air – 'I could measure it exactly, and it would be exactly 2,000 km long!' 'Don't be silly,' responded her father. 'Get on with your homework,' said her mother.

Who was right?

33

COVERING THE CHESSBOARD

On the left is a chessboard and a supply of dominoes. Each domino is large enough to cover two adjacent squares of the board exactly, and it is easy to see how you can place 32 dominoes to cover all 64 squares of the board.

On the right is the same chessboard, except that the top-left and bottom-right squares have been removed. Is it still possible to cover the board with dominoes, this time using only 31, so that every square is covered exactly once?

34

BOYS ONLY

On the island of Tutu, half the population was male and half was female. The King wanted to boost the male population. 'From now on,' he decreed to his people, 'you may have as many children in your family as you wish *until* you have a girl. Once you have had a girl, you will not be allowed to have any more children.'

Twenty years later, the King carried out a census. What had happened to the proportion of males on the island?

35

ELEMENTARY?

'Watson, I want to demonstrate one of the fundamental principles of deduction. Mrs Hudson, would you join us please. I have three handkerchiefs here: two of them are white, one is blue. Now both of you shut your eyes while I put one handkerchief on each of your heads. I will hide the third handkerchief so you don't know which one it is. Then I want you to deduce which colour is on your head.' They did as they were told. 'Right, Mrs Hudson, open your eyes. Can you tell me what handkerchief is on your head?' She shook her head.

'Well now Watson, from what Mrs Hudson said, I'm sure you can tell me your colour?'

'But she didn't say anything, Holmes.'

'Exactly.'

What colour was Watson wearing?

36

THE JUMPING FROGS*

On the left are three black frogs and on the right three white frogs. In between is a single space. All you have to do is to make the black and white frogs change places by moving and jumping them over each other.

You can only slide a frog into a space next to it, and you can only jump a frog over *one* other frog into the space beyond it. How can the frogs change places in as few moves as possible?

THE BRIDGE OF KNIVES

Take three table knives and place three tumblers on a table about one-and-a-half knives' lengths apart, at the corners of a triangle.

Then challenge someone to build a bridge between the tumblers (using only the three knives) that is strong enough to support the weight of a fourth tumbler placed in the middle.

WHERE IS THE EXTRA SQUARE?

Here are three squares made from twelve matches. How can you move just three matches to create five squares?

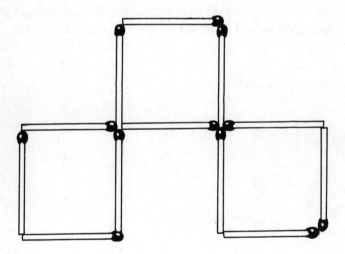

39

ROTATED CLOCK

Here is a clock. As you can see, all of the numbers have been removed. There is one other catch. The clock isn't necessarily the right way up. Is it possible to tell what the correct time is?

40

TAKING A TUMBLE

Jack and Jill were climbing on top of the garden shed, which was strictly forbidden because it was not very strong. Sure enough, a spar broke and they fell through the roof onto the sacks below. Jill got a large smudge on her face, but it was Jack who immediately went to wash his face. Why?

41

TUNNEL WIRES*

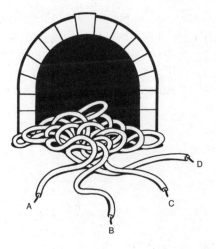

Engineers have just dug a tunnel under the River Spoog, and Bob Truffles has laid four electric cables along the tunnel. Unfortunately, the four cables look absolutely identical, and there are so many twists and turns in them that it is now impossible to tell which is which. He could walk through the tunnel untwisting the cables, but Bob has thought of a better way of sorting them out: his plan is to join two cables together at one end of the tunnel, starting with A and B, and then at the other end see which two cables make a complete electrical circuit when he puts power through them. The first test he does will obviously only tell him that two cables are either 'A and B' or 'B and A'. Using this method, how many trips along the tunnel does he need to make in order to label all four cables correctly?

42

HEADS AND TAILS

Linda arranged 16 coins in a square, alternately placed showing heads and tails, as in the figure below. 'All you have to do is to rearrange the coins so that the rows are all heads in the first row, all tails in the next, all heads in the third and all tails in the last row.'

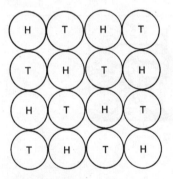

'What's the catch?' asked David.

'Oh yes,' said Linda, 'I almost forgot! You have to do it by touching only *two* of the coins.'

David pondered for a moment. 'Is it a trick?'

'Hmm, not really. Think about my rules.'

What is the neatest solution?

43

THE FARMER'S WILL

When Farmer Jones died, he left the quarter of his land that contained the farm buildings to his wife, and the remaining L-shaped area was to be divided equally between his four sons on the condition that the division would be into four parts that were identical in size and also identical in shape.

If his sons failed to divide the land according to this condition, the whole farm would go to Farmer Jones's wife. How should the sons have divided the land?

44

ONE HUNDRED UP

Peter and Mary were playing a game. They took it in turns to start, they took turns to play, and at each turn they named a whole number from one to ten, inclusive, and kept a running total of the numbers they had chosen. The loser was the first person to reach 100 or more.

Peter started the first game by calling 'seven'. What number should Mary have called to make sure she won the game?

45

A BARGAIN

'How much is one please?'
  'One pound.'
  'How much will two be?'
  'That will also be one pound, madam.'
  'I see, then I'll take 12.'
  'Certainly, that will be two pounds.'

What is the shop assistant selling?

46

SIX TUMBLERS

Tom was playing with six tumblers. He had filled three of them with water and the other three were empty. 'Look, they are full and empty one after the other,' he said, 'but I am going to make them so that there are three full ones, followed by three empty ones.' At this point his mother was convinced that he would spill them, so she took them away, but in fact Tom had a very simple plan to make three full next to three empty. What was it?

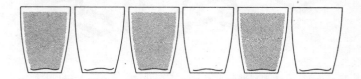

LINKING THE CHAINS

The blacksmith had five pieces of chain, each made up of three links. He wanted to join them all together in one long chain, so he broke one link in each of the first four chains, inserted a link from the next chain, and then sealed the link.

He was just finishing when his wife passed by. Seeing what he was doing, she pointed out that by breaking and joining four links, he had done more work than was necessary. How should he have joined the chains with as little work as possible?

48

NEXT LETTER, PLEASE

'What is it? Jumbled letters to make a word?' asked David looking at the piece of paper Anne was holding.

'No,' explained Anne, 'it's a sequence. I am trying to work out what letter comes next. It's all supposed to be logical and simple, but I can't see it!'

Can you work out what the next letter should be?

OTTFFSSE

49

CAN YOU BELIEVE
YOUR EYES?

Which of these two shapes is the larger in area?

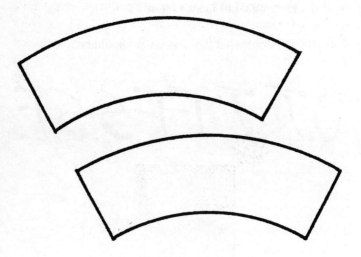

50

A TALE OF TWO SWIMMING POOLS

Fabian fforbes had a swimming pool in his back garden. His neighbours thought this was extremely posh, but he wasn't satisfied. He wanted to have a square pool that was double the size of the existing one, despite the fact that there were large and valuable trees at the corners, which the local council insisted he could not cut down. He did indeed end up with a square pool, double the area of his original pool, and all four trees remained in their places. How did he do it?

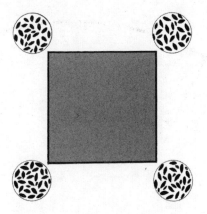

51

OUT OF THE SHADOWS

Have you noticed how the silhouette of an object sometimes doesn't look anything like the object? Here are three silhouettes of the same object, depending on how it is held against the light. What is it?

52

A RIDDLE PARTY

'Ooh, this is creepy!' said Cathy, reading out loud. 'The man who made it didn't want it, the man who bought it didn't use it, and the man who used it never saw it. What is it?'

'Here's another,' said Barry. 'You can only hold me for seconds, sometimes you lose me, and yet I am with you all your life. What am I?'

'What belongs to you which you would never want to get rid of, but which other people use far more than you do?' asked Peter joining in.

'What will grow if you feed it but dies if you give it water?'

'The more you take, the more you leave behind. What are they?'

Can you say?

53

A SAW POINT

'Here is a large cube,' said Sue, 'which I could cut with a saw into 27 identical smaller cubes like that.' She pointed to the cube on the table. 'The question is,' she continued, 'how many saw cuts would I have to make?'

'That's easy,' replied Martin, glancing at the cube. 'You make two vertical cuts, and then two more vertical cuts in the other direction, and then two horizontal cuts – that's six cuts in all.'

'Oh yes, of course,' said Sue. 'I forgot to mention that after you have made a cut, you are allowed to rearrange the pieces any way you like, before you make the next cut. So you might be able to use fewer cuts!'

Can you? What is the smallest number of cuts needed?

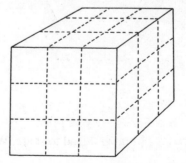

54

PAM'S PARTY*

Pam and five of her friends were looking forward to tucking into her birthday cake, which conveniently had six sides, when who should turn up but her cousin, Louise, who of course expected to join in the party and have her share of the cake.

'That's all right,' said Pam's mother, 'I can divide the cake like this, which makes nine slices, and Dad and I will have a slice each.'

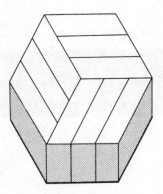

'But we shan't get so much!' complained Pam, 'there must be a fairer way.' Indeed, there was a way to cut the cake into just eight identically shaped pieces, one each for the children and one for Mum and Dad to share. What was it?

55

CAN YOU SEE STRAIGHT?

One of the two lower lines is the continuation of the line above the bar. Which is it?

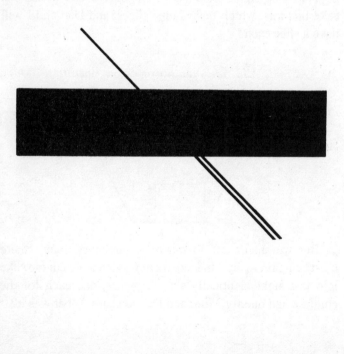

56

MENIAL COPYING*

'Stephen, sorry to dump this on you, but could you possibly take photocopies of this marketing strategy document, personally deliver one copy to everyone who is coming to my management meeting at noon, and return the original to me.'

'But it's after half past eleven now, Brian.'

'Yes I know, I'm sorry to be a pain, must dash.'

Stephen, efficient as ever, did meet the deadline, though 159 sheets of paper were churned out of the photocopier in this exercise.

How many people do you think went to Brian's meeting?

TURN FOUR INTO FIVE*

It is not difficult to get from TWO to SIX by changing one letter at a time, each time making a proper word, like this:

TWO
TOO
TOP
TIP
SIP
SIX

So TWO to SIX can be done in five steps. FOUR to FIVE takes longer. How many steps do you need to turn FOUR into FIVE? Can you do it with every letter changing at least once, including the F? How about turning ONE into TWO?

58

A COLUMN CONUNDRUM

Write the numbers 1 to 9 on pieces of card, and arrange them in two columns like this:

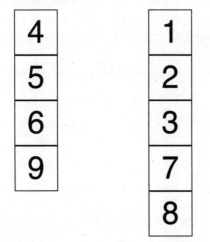

How can you get the two columns to add to the same total by moving just one of the cards?

LUCKY FRIENDS

'The aunt of somebody in my class won £1,000 in the National Lottery,' said David at supper that evening. 'Well somebody at our school's family won £10,000,' said his sister. 'Funny that, one of my friends at work was telling me the other day that a friend of his had a decent win,' said their father. 'Gosh we do know some lucky people don't we – it seems that everyone is lucky except for us,' said mother.

How do you account for them having so many lucky friends?

60

QUICK THINKING

A bottle and its cork cost £1.10 and the bottle costs £1 more than the cork. How much does the cork cost?

61

IS THIS THE WORLD'S SIMPLEST GAME?

This game is so simple that when a group of primary school pupils were shown it a while back, they quickly discovered the winning strategy and then decided that it was no longer worth playing. Can you do the same?

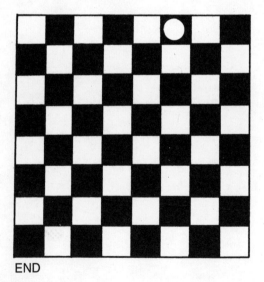

END

The rules are extremely simple. The single piece is placed somewhere on the board, and then the two players take it in turns to move it from the square it is on to any new square, so long as they move it in a straight line parallel to the sides of the board. (In other words, like a rook in chess.) Every move, however, must be either down the board, towards the bottom edge, or leftwards, towards the left-hand edge. The winning player is the one who makes the final move on to the square marked END.

We show a typical starting position. It is your turn to play. What move do you play to make your opponent squeal?

62

SCREW ON YOUR THINKING CAP!

The picture shows one hand, on the right, turning a bolt in a clockwise direction, as if screwing it into a nut, and another hand, on the left, turning another bolt anti-clockwise, as if unscrewing it.

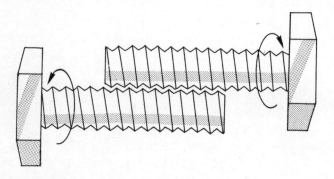

As it happens, the two bolts are in close contact all the time, so they will either be drawn closer together, or will be pushed further apart, or, perhaps, not move closer or further apart. Without getting a pair of bolts to experiment with, can you predict the correct answer?

63

SPOTTING DOOBREYS

Edward Spinks is a keen bird watcher, so last week he was particularly excited to discover that a pair of Lesser Wattled Doobreys are nesting in the tree at the back of his garden. He has set up a hide in his back room, and every evening when he returns from work he sits down to study the birds. He has observed that the Doobreys have young chicks, and in order to feed them, the male and female are on a non-stop food search. The two parents work separately, flying off over the hedge, returning about fifteen minutes later with grub (literally) and then within seconds flying off again on their next search. But there is something that Edward cannot explain. His first sighting of a bird every day is sometimes the male and sometimes the female. But the first sighting has always been of it flying back to the nest, rather than leaving the nest. Even when he gets home early, he still doesn't see a Doobrey leave the nest until he has seen one fly back. Can you think of the most likely explanation for this interesting ornithological phenomenon?

64

TRUE OR FALSE

In the list below, which of the statements is true, and which is false?

1. The number of false statements in this box is one.

2. The number of false statements in this box is two.

3. The number of false statements in this box is three.

4. The number of false statements in this box is four.

65

LORD HENRY'S HUNT

The fox hunters were gathering outside the Red Lion. It was a hot sticky day and several of the hounds were slouched on the ground having a doze before the off. Lord Henry was getting changed while planning the after-hunt celebrations with his butler, Basil. 'Oh, by the way, pack my box with five dozen liquor jugs,' Henry was saying. Then suddenly, out of the blue, a creature shot out of the bushes and bolted off down the road. 'Excuse me,' said Basil, 'but did you see that? A quick brown fox jumped over that lazy dog!' 'Action stations,' yelled Henry. 'Quick, Baz, get my woven flax jodhpurs!' Basil fetched the special trousers, Henry squeezed himself in, and with a call of 'Tally ho!' he led the hunt off into the neighbouring field.

Meanwhile, did you notice more than one interesting feature in what they said?

66

CROSSING THE MOAT

You have discovered an old derclict house, with a moat around it which is still full of water and about 4 metres across. Naturally, you want to get to the house, but even if you could take a run up and jump across the moat from this side, you would not be able to jump back.

However, you find two planks of wood which, when they are lowered across the moat, very nearly reach to the opposite bank – they are just a few centimetres short. You would like to tie the planks together but you have nothing with which to tie them securely. How can you reach the derelict house by using the planks, and without getting your feet wet or being in danger of falling into the water?

67

TRICKY ENDING

Using only one word, and correct grammar, complete the following sentence as accurately as you can:

'The number of occurrences of the number one in this sentence is . . .'

68

FAULTY SCORING

The scene is Ribblesfield Tennis Club, it's the final of the club championship no less, and John Tuttle and Bill Hemp are in a nail-biting game of singles. The electronic scoreboard shows the match at 5 games all.

[Sound effect] Pok pok pok
Umpire: 15-love
[Sound effect] Pok pok pok pok
Umpire: 15-all
[Sound effect] Pok
Umpire: 30-15
[Sound effect] Pok pok pok pok pok OUT!
Umpire: Deuce.
Tuttle: Umpire, you must be joking.
Hemp: Tuttle's right. Everyone knows deuce doesn't come after 30-15!
Umpire: Sorry, I know it's unusual, but let's leave it at deuce anyway, it's a bit hard to adjust the scoreboard now – unless you want to start this service game again.

After a heated discussion the players finally concede that the umpire leaving the score at deuce is quite fair while starting the game again would give Tuttle an unfair advantage. How do you explain the umpire's decision?

69

A SIX PACK*

Fitting four words into a word square, so that the same words read across as down, is not difficult. But can you fit six words into a four-by-four word square?

For example, can you fit these words: AREA, REAR, DEED, DART, BRAT, BARD?

70

CHRISTMAS BIRTHDAYS

At the Hetherington family gathering over Christmas, Dad commented: 'Something strange seems to be happening to our family's birthdays. Have you noticed how New Year's Day comes exactly one week after Christmas Day? Yet the year Mum was born, Christmas was on a Tuesday and New Year was on a Monday. And Damien, two days ago you were eight years old, yet next year you will be eleven. And Janet seems to be extremely lucky. Even though she was also born soon after Christmas, her birthday is always in the middle of summer.'

The family groaned. What date was it, and how do you explain the strange birthdays?

71

CHOPPING THE CHOCOLATE

Barry was the happy owner of a giant chocolate bar, which he wanted to break up into 35 small squares. However, he couldn't decide how to do it. He thought of breaking it into five strips and then breaking up the strips, but then he wondered whether he should break it instead into two large rectangles, and break up each of these separately.

Naturally, he wanted to make as few breaks as possible. Can you say what was the smallest number of breaks needed?

72

POLAR BEARINGS

You are travelling to the North Pole by sled, but your huskies are too enthusiastic and you overshoot the Pole slightly. Will East be to your left or right?

73

SCATTERED VEGETABLES

There is a very logical reason why one of the list of vegetables below is the most appropriate to insert in the following sentence. Can you say which?

'I do not like those people leaving _____ scattered everywhere.'

(a) sprouts
(b) cauliflowers
(c) potatoes
(d) swedes
(e) carrots

A BUCKETFUL OF TROUBLE

You are in possession of two plain cylindrical containers, as the illustration shows. One will hold 3 litres, and the other 7 litres.

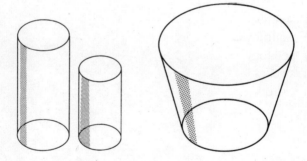

To start with they are both empty, but you have an endless supply of water from a tap. How can you leave 15 litres in the bucket on the right, in just four moves?

75

SUPERBALL REBOUND

Craig has a superball – so super that when you drop it, it bounces right back to the height you dropped it from. He's wondering about what would happen if he threw the ball at something moving. Suppose he was standing on a railway line and a diesel train was bearing down on him at 50 mph (don't try this at home!). Suppose he then threw the superball at 20 mph straight at the oncoming train. When the ball bounced off the front of the train, how fast would the ball be coming towards Craig?

76

A STANDARD ANSWER?

What comes next in this sequence, in which all the question marks represent the same thing?

ST   ND   RD   ?   ?   ?   ?

77

SWITCHING OLIVES

Tim's mother has meticulously sorted the nibbles for tonight's party so that there are 100 black olives in the black dish, and 100 green olives in the green dish. They look beautiful. Tim, however, thinks this is boring, so unknown to his mother he takes twenty green olives and puts them in the black dish. Then he mixes up the black dish and takes twenty of the olives in this mixture and puts them back in the green dish.

His mother is furious. 'Now I'm going to have to sort them out. I don't know whether there are more green olives in the black dish, or black olives in the green dish.'

Can you help?

78

THE ALLOTMENT PLOT

Mr Brown, Mr Jones and Mr Smith each had two plots on the local allotment, as this map shows. One day, Mr Brown announced, 'I've decided to build a path straight from one of my plots to the other.'

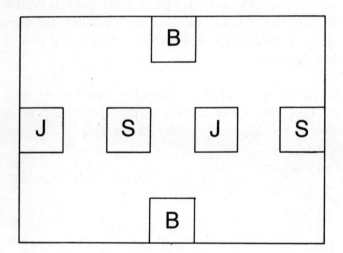

'You can't do that!' exclaimed Mr Jones. 'It will be in the way when I want to go between my two plots, and anyway, it will stop me building my own path!'

Mr Brown wasn't interested in arguments and stubbornly refused to change his mind. So Mr Smith suggested to Mr Jones that they get in first and build their own paths, before Mr Brown had a chance to lay his down.

They did so, and when Mr Brown came to build his path, he discovered that he had to pay for his selfishness.

How were the three pairs of plots joined by paths that did not cross each other?

79

LEAVE AND LET DIE

'So Missterrr Bond, now zat I haf shown you how I plan to take over ze vurld, it is time for me to put you to a slow and agonising death. I haf stripped you of your shirt and tied you so you cannot move. I am going to seal you into zis air-tight chamber, vith barely enough room to sving a cat. It is so vell insulated zat no heat can enter or leave it. In ze chamber you vill notice I have left a large kitchen freezer svitched on at full power – and vith ze door jammed open! Already zis chamber is at only 5 degrees Centigrade. Goodbye and say your prayers, Mr Bond.' Roughly how long do you reckon that James has to find his escape before he freezes to death: minutes, hours or days?

80

AMAZING EQUATION

'Did you know that twelve plus one is the same as eleven plus two?' asked Claire.

'Yes, and . . . ?' said Nicola.

'I'm not just talking about the numbers. I can prove it using just these wooden blocks,' Claire added.

On each of the nine blocks was a single letter. What letters were they?

81

T-SHIRT TEASER

Scruffy Sam has put on his T-shirt. Unfortunately, it is inside out and back to front. Normally the label is on the inside at the back. Where is it now?

82

SECRET NUMBER CLUB*

On his first day at college, maths student Derek found the following curious note in his pigeonhole:

'I belong to a secret code club. Each of us has a special whole number (between 6 and 9) and before we write down any number we always add our own special number to it. There are 10 members in the club, and there are 6 others who have smaller numbers than me, and 5 with bigger numbers than me. All of our special numbers add up to 27. If you can tell me my number and those of the other members, you can join the club. Yours, Archie Meades.'

Derek was convinced that there was a mistake, but he was wrong. Can you help him?

83

SURPRISE MEAL

'Darling, I've seen so little of you recently that one evening this week I am going to take you out for a meal, but the evening I choose will be a surprise,' said Andrew on Monday morning. 'That would be nice,' said Jenny, who immediately tried to work out which night Andrew would choose.

'Well, it can't be Friday,' she thought, 'because then by the time we get to Friday morning and we haven't been out I'll know it must be that evening, so it won't be a surprise. But since it can't be Friday, then that means it can't be Thursday either, because on Thursday morning I'll know it's going to have to be on Thursday evening. But if it can't be Thursday or Friday then by the same reasoning Wednesday can't be a surprise either . . . and it can't be Tuesday . . . or tonight . . . He's not going to take me out for a surprise meal at all, the lying so and so. How typical!' Her logic seemed impeccable.

But when she got back to the flat that evening there was a message on the answerphone from Andrew: 'Jenny, about that meal: how about going to the Tandoori restaurant tonight?'

Jenny was surprised. What had been wrong with her reasoning?

84

FIVE COINS IN CONTACT

'Can you place four coins so that each coin touches the other three?' asked Peter.

'I certainly can,' replied his uncle, 'and if that isn't enough,' he continued, 'I can place five coins so that they are all touching each other!' Peter was silent, thinking.

Can you discover how to place four coins so that each touches each of the others, and discover a way of placing five coins under the same conditions?

85

SPOT THE DIFFERENCE

The leopard and the cheetah wanted to find out which of them was fastest. 'That little stream is 100 paces away, I'll race you to it,' said the cheetah. Off they ran, and the cheetah was ten paces ahead of the leopard when she sprang over the stream. 'I'm too fast for you,' said the cheetah. 'I tell you what, let's do that race again, but this time you start from the same place, and I will start ten paces behind you.' The leopard agreed, and they ran the race again at exactly the same speeds. Who won this time?

86

SET IN ORDER *

Mr Knowall's children have been using his favourite set of encyclopaedias for their homework research, but as usual they have not bothered to put them back in the correct order. In fact they are horribly jumbled.

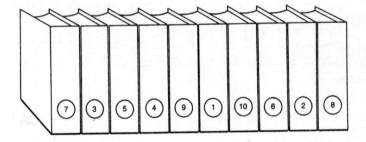

Mr Knowall wants to put them back in order, but naturally, because each volume is very heavy, he wants to take as few volumes as possible off the shelf. If one move consists of taking a volume off the shelf, pushing some of the remaining books to one side and replacing the volume he took off, what is the smallest number of moves he requires?

87

THE DRAUGHTSMAN'S POSER

Here are the front, top and side views of an object. Can you sketch it, or describe it?

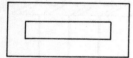

88

PARDON ME, BOYS

Mike and Merv work as shoe shiners outside the station. To shine the shoes, they first apply the polish (which takes one minute) and then they rub it up to a shine (a three minute job). The last train leaves in just under eight minutes and they are about to finish for the day when three customers arrive. Can Mike and Merv shine the shoes and still both catch the train?

89

OLD MACDONALD

Old MacDonald had a farm. And on that farm he had . . .

One animal with half as many letters now as it had when it was younger.

One animal with fewer letters now than it will have when it is older.

One animal with half as many letters as its plural.

One animal with the same number of letters as its plural.

And altogether the vowels they had were E-I-E-I-O (though not in that order). What were the animals?

90

HOWZAT?!

Sam turned up to watch his dad playing cricket, but since watching dad batting is far from interesting, he got on his bike and decided to do a circuit of the field, going anti-clockwise as indicated by the arrows.

After five minutes he arrived back at his starting point. What is interesting is that in his whole journey he never once turned left. How did he do it?

91

TO AND FRO

Mrs Smith takes on average one hour to drive to work, and on average she overtakes 8 buses going in her own direction (all no.10s, which is the only service on her route), and passes 16 buses coming the other way.

How often do the buses run?

92

QUIRKY QWERTY

When the typewriter was invented, there was a problem with the keys jamming with each other. To reduce the chance of this, the proprietor of the first patent mixed up the keyboard letters to slow down the typist. This is why the top row has the letters QWERTYUIOP. Now that we have computers, it would make sense to have more sensible lettering, but the world has invested so much money in QWERTYUIOP that it looks like we are doomed to using it in perpetuity. But still, it does mean that a classic old puzzle still holds true. Can you find a ten-letter word that uses letters only from the top row of the typewriter keyboard (not necessarily all of them)? Even better, can you find three such words?

93

WHODUNNIT? *

Shortly before the trial of Charlie Snod, the police received a mysterious note from the infamous villain Henry van Eyck. It read as follows:

'Regarding the current case concerning Charlie Snod, here are six statements on the matter:

1. The truth is that this statement and the one after the first true statement are not both true.
2. If you add the number of the first false statement to the second true statement you get the number of a statement which is as true as number one.
3. Statement number 2 is a downright lie, I'm afraid.
4. Charlie Snod is innocent or I'm a Dutchman (or both).
5. At least half of these six statements are true.
6. My own gang was responsible for the crime.

Yours honestly,
Henry van Eyck.'

Amazingly, the logic in Henry's note turned out to be impeccable. Whodunnit?

94

BAD LINE

The telephone rings in the office:

Caller: Can I talk to Mr Jardine, please?
Secretary: What's the name, sir?
Caller: Paten.
Secretary: I'm sorry, it's a bad line. Can you repeat that.
Caller: Paten. P for Pluto, A for Adolf, T for Tummy.
Secretary: T for what, sir?
Caller: T for Tummy, E for Elephant, N for Nose.

According to the old puzzle, this conversation demon-
strates that the secretary is not very intelligent. Why? And
do you agree?

95

A BUN FIGHT*

Mrs Watson had just taken six hot and delicious buns from the oven for the children's tea, or so she thought. When she looked at them again, one of them had already gone. 'You're the limit!' she cried to Emma and Pippa, 'Now you won't get any more! I shall eat them all myself!'

Then she caught herself; she had no wish to put on more weight by eating five buns she didn't need. 'On second thoughts, you can do some extra homework! If you can work out how to divide the buns equally between you, without removing them from their tray, with one straight cut only, then you can have them for tea. Otherwise, Fido gets them!'

Emma and Pippa were successful, in fact they each found one way to divide the buns. How did they do it?

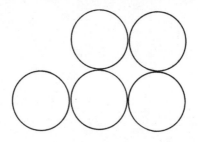

96

BEER MAT BET

'I will bet you a fiver,' said Gary to Marian, 'that if we take turns to place beer mats on this table top, and I start, then you will be the first one who is not able to place a mat on the table! Of course,' he added, 'the mats must not overlap each other, or fall off the edge of the table.'

The beer mats were all circular and identical in size, and the table top was a rectangle.

How did Gary plan to win this game?

97

THE GLASS THAT CHEERS*

It was time for the Christmas punch to be handed round, a sociable activity that Scrooge loathed, so every year he put it off for as long as possible. This year he had a tricky puzzle, using all the punch glasses, to offer his guests.

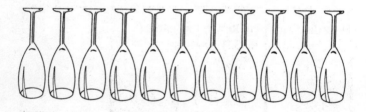

'Before we dip into the bowl,' he said to his assembled guests, 'I invite you to solve this puzzle. You have to turn all the glasses the right way up, by inverting three at a time. When you invert three glasses, that counts as one move. And you must do it in the smallest number of moves possible. How many? Eh?'

How could his guests get their punch?

98

ALF'S CORNET

Here are four cards. All of the cards have a picture of an ice cream on one side and a flavour on the other side. 'I think that every card with a lolly on one side has the word "chocolate" on the other side,' says Alf. Which cards do you need to turn over so that you can prove that what Alf says is true?

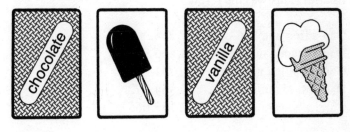

99

ADD UP TRICK

Charles decided to impress his father. 'Write down any random seven figure number please, then write down any other seven figure number under the first one.' His father did this.

'Now I will write one more number, and then I want you to write down another one, then I will write a second,' said Charles. This was the result:

Dad's first number –   7,258,391
Dad's second number –   1,866,934
Charles's first number –   2,741,608
Dad's third number –   5,964,372
Charles's second number –   8,133,065

'Right,' said Charles, 'I bet I can add these numbers up in less than five seconds.' 'Nonsense, even with a calculator you couldn't do it that quickly,' said his father. But in three seconds Charles had written down the answer to the sum, which was 25,964,370. How did he do it?

100

SHOP REFLECTION

Standing in the street in front of a shop window, I can see the name of the shop in large gilt letters on the window itself, and I can also see it reflected in a mirror inside the shop. Do I see it on the mirror the right way round, or reversed?

101

CATASTROPHE CAT

Unfortunately for Sybil, the officers' cat, she has chosen to spend the night asleep on the top of the caterpillar track of one of the large tanks parked outside the camp. It is unfortunate because at the crack of dawn there is a military exercise, and the tank starts up and drives off at its regulation 10 mph. The cat, clever enough to realise she might be on the verge of getting squashed but too sleepy to realise her best policy would be to jump off, instead starts running backwards along the caterpillar track. How fast will Sybil have to run to avoid a grim end?

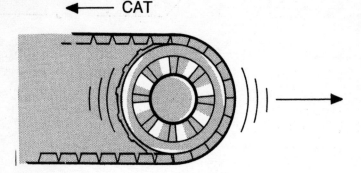

102

NAME PREDICTION

Peter Higgins was walking down the high street when he bumped into an old friend. 'Hello, I haven't seen or heard from you since graduation back in 1982!' said Peter. 'What's happened to you?'

'Well, I got married in 1989 to somebody you wouldn't know. This is our daughter,' said the friend, who was holding hands with a little girl.

'Hello, and what's your name?' said Peter to the girl.

'It's the same as mummy's.'

'Ah, so it's Jane is it!' said Peter.

How did Peter know?

103

ICE IS NICE*

'Who wants ice cream?' asked Mrs Smith. 'We all do,' said Peter, so Mrs Smith started to cut the cubical block into three equal slices in the obvious way. 'Oh, no! Boring!' cried Gary, 'We always have boring old blocks, we want something DIFFERENT.' 'Yes, let's have something new,' said William, who didn't notice that he was repeating what Gary had said.

Mrs Smith thought for while, and then sliced the cube of ice cream into three pieces, identical in shape, which used up the whole block, and looked nothing like the usual boring slices.

How did she do it?

(A word of advice: unless you have superpower ability to see things in your head, it is helpful to cut cubes from cheese and experiment – you can eat the cheese later.)

104

TIME TRAVEL

Some years ago the eccentric adventurer Bart Carruthers decided to fly his small plane single-handed around the world. He left London at noon on 1 March, and his route took him via Cairo, Hong Kong, Hawaii, New York and finally across the Atlantic. He arrived back in London at the end of his marathon journey at noon on 31 March. Bart kept no diary and so to keep track of time he simply counted the total number of nights he had over his journey. How many was that?

105

HEADS YOU WIN

You have four coins and I have three coins. We both toss all of our coins. What is the chance that you end up with *more* heads than I do?

(a)   Higher than 50–50
(b)   Lower than 50–50
(c)   Exactly 50–50

106

A DICEY CUBE

Mark has made a cube at school, and drawn some numbers on the faces. Here are three views of the cube. As you can see, one of the views shows only one face front on, the number 1.

What number is on the opposite face to 4?

A DICE GAME

ANSWERS

1. Solution:

The answer is *not* 'about four inches'. In fact, the bookmarks are almost touching.

This picture shows how the two volumes look when they are placed on the shelf.

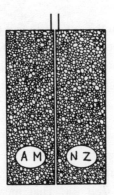

The bookmarks are placed at the front of Volume I and the back of Volume II. Since *aardvark* is near the front of Volume I and *zebra* is near the back of Volume II, the bookmarks are next to each other, separated by little more than the covers of the two volumes.

2. Solution:

Martin was wrong. The rolling coin makes a complete turn even though it has only travelled around half a circle. In the diagram, the quarter circle AB is the same as AC, so by the time B rolls round to C the coin will already be upside down after completing only half its journey. By the time point D rolls round to point E, the coin will be the right way up again.

3. Solution:

The missing word in this quiz is in fact the word 'quiz'. The six words collectively use the 26 letters of the alphabet. (Collections of words that between them contain all the letters are known as pangrams.)

4. Solution:

She opened the *Mixed* tin. Suppose she took out a *Rich Tea* biscuit from that tin. Since the label was wrong, that must really be the *Rich Tea* tin. That means the tin labelled *Digestive* can only contain *Mixed* biscuits, and the one labelled *Rich Tea* must therefore contain the *Digestive* biscuits. Just the same logic applies if she removes a *Digestive* from the *Mixed* tin.

5. Solution:

Mr Jones wrapped the cue diagonally to make a package that, when measured end to end, was just under 1.5 metres in length.

Of course, when measured diagonally the parcel was as long as a cue, but the Post Office regulations did not say anything about diagonal measurements.

6. Solution:

This feat can be achieved by moving two matches as shown, so that the cherry appears inside the glass.

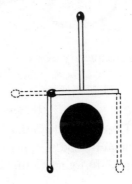

7. Solution:

This is a true story. There was a special deal: buy 10 cards and get 1 (or 2) free. This means nobody would buy 10 since they might as well take the extra 2 for the same price. The deal was 50p per card or 12 for £5. So while buying 5, 15 or 20 cards was cheaper than buying 6, 16 or 21, buying 10 was not cheaper than buying 11, or indeed 12!

8. Solution:

It is not knotted. Pull the right hand end and it first turns into this . . .

and then into this . . .

and finally becomes completely free.

9. Solution:

Sue sent the letter. Every time the letter 'z' is printed, the daisy wheel shifts round by one. Z appears six times in the message, so the wheel shifts round six times in total. The rest of the message reads:

Every time I type the letter z it starts to go funny. I must get it fixed when I come home. Lots of love, Sue

10. Solution:

Mrs P put down £1 in small change (20 pence and 10 pence coins), from which the driver could see that she wanted a £1 ticket because otherwise she would have given him exactly 70p. Mrs Q put down a £1 coin, which could be used to pay for either type of ticket.

11. Solution:

Maria filled her can at the spot marked Z. Remember the old principle that the shortest distance between two points is a straight line? That principle applies here too – but it needs a mirror.

Imagine that the second half of her return journey, from Z to X, is reflected in the line of the river bank, so that the reflection of X is at P.

Then the total length of Maria's journey is from Y to Z to P and this will be shortest when YZP is a straight line. In other words, when the angle between YZ and the river bank is the same as the angle between XZ and the river bank.

So Maria ran to the river bank and 'bounced' back to the picnic spot, like a snooker ball bouncing off a cushion.

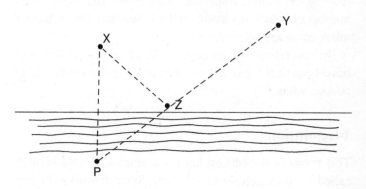

12. Solution:

It does not matter how fast she runs – even a world record sprint could not increase her average speed to 6 mph. This is because the Fun Run is 3 miles long, so if her overall speed was 6 mph she would do the whole run in 30 minutes. But she has already taken 30 minutes to run the first 2 miles at 4 mph, so she would have to run the final mile in 0 minutes! (This is a good illustration of the fact that averages can be quite complicated. You usually can't average an average.)

13. Solution:

The only place on Earth from which every direction is north is the South Pole, so that is where Wimpish must have been.

Due to the curved surface of the Earth, if you start from the North Pole and walk south, due east and then due north, you will return to the North Pole.

You can perform the same trick if you start a little over ten miles north of the south pole, walk ten miles south, then walk ten miles in a circle round the South Pole – making an exact number of complete circuits of the Pole – and then return ten miles north to your starting point.

However, there are no bears in Antarctica, so Tuffer must have been at the North Pole, where the polar bears are, of course, white.

14. Solution:

This is one of the earliest known examples of what is often called a 'think of a number' trick. Since it works for any number, let's call that number 'Blob'. Pick up 'Blob' beans in each hand. Now follow the instructions:

	Left hand	**Right hand**	**Total**
Start	Blob	Blob	Two Blobs
Four across	(Blob + 4)	(Blob – 4)	Still two Blobs
Cast away all in right	(Blob + 4)	0	Blob + 4
Do same no. from left	8*	0	8
Pick up five more	8	5	13

* (Blob + 4) take away (Blob – 4) equals 8, whatever 'Blob' is.

In fact, this use of a symbol ('Blob') to represent any number is an example of algebra. Instead of calling it 'Blob', mathematicians usually prefer 'x'.

15. Solution:

Make an ordinary knot in a strip of paper, and very carefully flatten the knot, keeping it tight as you do so. It will form a regular pentagon.

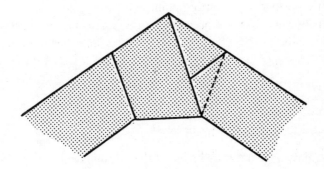

16. Solution:

The man took the goat across, leaving the wolf and cabbages behind. Then he returned and took the wolf across and brought the goat back. He then rowed across with the cabbages. Finally, he returned to pick up the goat and take it across, making a total of seven crossings.

17. Solution:

After 280 seconds, the Big Wheel would have gone round exactly 28 times, the Rotator would have gone round exactly 35 times, and the Twirler would have gone round exactly 40 times.

This happens because 280 is exactly divisible by 10, 8 and 7; it is also the smallest number with that property. So they were in line again after 280 seconds, or just under 5 minutes.

18. Solution:

Move the bottom left toothpick as shown.

You now have a mirror image of the original deer.

19. Solution:

There were 20 crows to start with. Each time Farmer Giles shot at them, half of them (10) flew away, but the same number soon returned.

20. Solution:

Dougal's original pocket money was £6 per week. Compare the two ways of increasing his pocket money. This tells you that one third of the original, plus £1, is equal to one quarter of the original, plus £1.50.

The difference between one third and one quarter is one twelfth, so the extra 50p is one twelfth of the original pocket money, which must therefore have been 50p x 12 = £6.

21. Solution:

The carpenter made this cut through the original plank, as in the first figure, and then rearranged the pieces as in the second illustration.

22. Solution:

Forty. (In French there are several numbers: *deux*, *cinq*, *dix* and *cent*.)

23. Solution:

The trains up and down the line arrived every ten minutes, but the down train was always two minutes after the up train. So he was four times as likely to arrive in the eight minutes before the up train arrived as he was to arrive in the mere two minutes before the down train arrived.

24. Solution:

The word that becomes decoded is 'DECODED': it is unchanged when you look at it upside down in the mirror (have a look and see!). The reason is that all the letters in that word are symmetrical top and bottom. All the other words contain at least one letter that does not have this line of symmetry.

25. Solution:

Debbie is a small girl going to school and back. When she enters the lift in the morning she can reach the bottom button marked 'GROUND' but, on returning, she cannot reach any button higher than 'FLOOR 5'. Only if an adult is getting into the lift at the same time can she ask them to press 'FLOOR 16' for her and go all the way up. (This puzzle has traditionally been presented as being a dwarf in the lift, but it has always seemed absurd even in the fantasy world of puzzles that it never occurred to the dwarf to either (a) take a stick, or (b) move apartment.)

26. Solution:

Harry Hoppit was setting out for the Antipodes Islands, east of New Zealand, so called because they are on the opposite point on the globe to London, so it made no difference in which direction he set out – he would have to travel halfway round the world to get there.

27. Solution:

No, he wasn't. He handed over his £10, and Janet said, 'Thanks,' and started to walk away. Her brother said, 'Hey, if you don't give me £15, you've lost the bet!' 'Oh, yes, so I have,' said Janet. 'So I owe you five pounds.' She handed £5 over and went off to the cinema with £10 in her pocket and a smile on her face.

28. Solution:

When he walks up the escalator, Mr Watson takes 20 seconds to reach the top. In that time he has walked up 40 steps (two steps per second). If he stands still, in 20 seconds he only rises by half the height of the escalator. So the height of the escalator, H, is 40 steps plus half of its height, H, which means the escalator must be 80 steps high.

(This is similar reasoning to the old puzzle 'If a brick weighs one pound plus half a brick, how much does a brick weigh?' – to which the answer is 'two pounds'.)

29. Solution:

These are the moves the trains should have made.

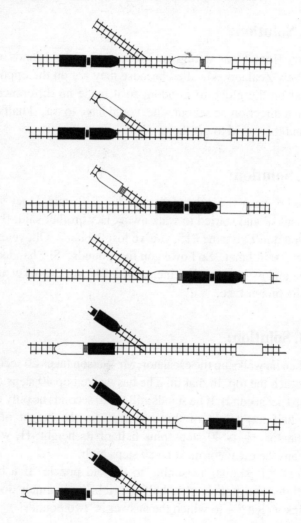

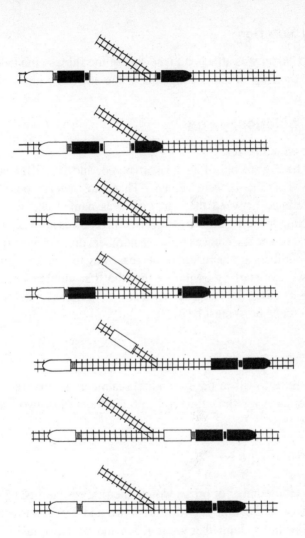

30. Solution:

The ladder was attached to the side of the ship, so the ladder and the ship would rise together as the tide came in.

31. Solution:

She won 5 prizes.

There were only four girls involved, and six differences between them were given. Therefore every possible difference between the four girls was named and we are looking for four numbers whose six differences are 1, 2, 3, 4, 5 and 6. The differences between numbers don't change if all the numbers are increased or decreased by the same amount, so we can start by assuming that the first number is 1, in which case the largest number (to give a maximum difference of 6) must be 7.

1 __ __ 7

We have to fill in the two middle numbers so that the six differences are the numbers 1 to 6. There are just two ways to do this:

1 2 5 7 or 1 3 6 7

The total number of prizes won in the first case is 1+2+5+7 = 15, which can be raised to 27 if we add 3 prizes to each girl's total, making their totals 4, 5, 8 and 10. The total of the second set is, however, 17, which cannot be raised to 27 by increasing each number by the same amount. Therefore

4, 5, 8, 10 are their numbers of prizes, and Bernice, who won 1 more than another girl, won 5 prizes.

32. Solution:

In a way they are all right! Rivers make up part of the Spain–Portugal border. A mathematician called Hugo Steinhaus pointed out in 1954 that the length of a river depends on the size of the ruler that you use to measure it. In particular, he pointed out, if you use a small ruler and trace its banks in and out of every little creek and inlet, every twist and turn, then you will end up with a total length far in excess of the length calculated from a map, or given in geography books.

So when the Spanish claimed years ago (yes, they really did) that their boundary with Portugal was 987 km, they were probably correct, by the scale they used to measure it, and the Portuguese may well have been correct when they claimed about the same time that it was 1,214 km. Pamela is most likely to be wrong, because if she really tried to get down on her hands and knees and measured the length of the border with a school ruler she would end up with a total length far more than 2,000 km.

33. Solution:

No, it is *not* possible! However you place the 31 dominoes, there will always be two squares left that are not adjacent to each other, as the diagram overleaf illustrates. Trial and error will also show that the two squares that remain are always *both* black squares. This is a clue to the explanation. When a domino is placed on the board it covers two adjacent squares

– one white square and one black square. However many dominoes you place, they will always cover equal numbers of black and white squares.

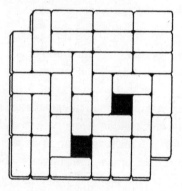

But the board with the corners missing has 32 black squares and only 30 white squares so, after placing 30 dominoes, two black squares must remain, and these can never be covered by the 31st domino.

34. Solution:

The proportion of males on the island was still half! The King's policy had no effect at all on the ratio of boy babies to girl babies. Telling parents to stop having children when their first girl is born is the same as asking them to stop tossing a coin when the first head appears. On average there will still be 50% boys and 50% girls. (If you don't believe us, toss a coin to simulate the King's rule.)

35. Solution:

Watson was wearing a white handkerchief. If he had been wearing the blue handkerchief, Mrs Hudson would immediately have known that hers was one of the two white handkerchiefs. The fact that she didn't know means that she saw a white handkerchief on Watson's head.

This type of puzzle illustrates the interesting principle that sometimes you can make useful deductions from apparently 'no' information. There is a famous example from a real Sherlock Holmes story called 'Silver Blaze'. Inspector Gregory asks whether there is anything which Holmes wishes to draw to his attention.

'To the curious incident of the dog in the night time,' [replied Holmes]. 'But the dog did nothing in the night time.' 'That was the curious incident,' remarked Sherlock Holmes.

36. Solution:

Number the squares from 1 to 7, as in this illustration.

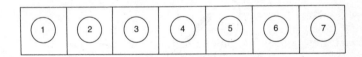

Then you need to make 6 slides and 9 jumps, a total of 15 moves. This is how it is done: 3–4; 5–3; 6–5; 4–6; 2–4; 1–2; 3–1; 5–3; 7–5; 6–7; 4–6; 2–4; 3–2; 5–3; 4–5.

37. Solution:

Place the knives so that each knife goes over the knife to its left and under the knife to its right, or the other way round. This three-legged bridge will not collapse, and will take a considerable weight – much greater than the weight of one tumbler.

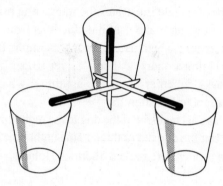

With four knives, you can afford to move the four tumblers even further apart.

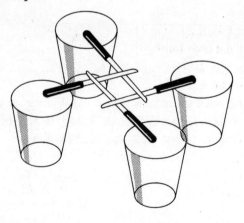

38. Solution:

Moving these three matches leaves four small squares and the large outer square.

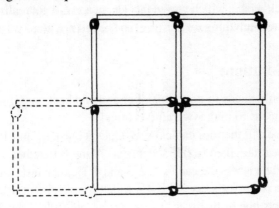

39. Solution:

Yes. There is only one position in which the little hand is pointing in the correct direction. If it is quarter past the hour, the little hand should be quarter of the way between the hour marks, if it is twenty past the hand should be a third of the way, and so on. Since the hand is exactly midway between two of the hour symbols, it is half past the hour . . . which makes it 1.30.

40. Solution:

Jack saw that Jill's face was dirty and assumed that his face was also dirty, so he went to wash it. Jill did not see anything on Jack's face, which was in fact clean, and did not realise that there was anything on her face, so she did not go to wash it.

41. Solution:

Bob only needs to go to the other end and back once, and he only needs to make two connections!

If we call the four cables A, B, C, D at one end, and 1, 2, 3, 4 at the other, then Bob first connects A and B together. At the far end he labels the cables 1, 2, 3 and 4. He then tests until he finds a pair – let's say 2 and 4 – which form a circuit. He now connects one of that pair to another unused cable – say 2 to 3 – and goes back to the other end of the tunnel, disconnects A and B and again looks for the pair which form a circuit. One of these will be A or B (the other end of 2), the other will be C or D (the other end of 3). Let's suppose the pair is A and C. He now has all the information he needs. In this example:

A is 2, so B is 4, C is 3 and D must be 1.

Of course this works whichever combinations turn out to form the circuits.

42. Solution:

Place your fingers on the two heads in the bottom row and slide them round to a position exactly above the two tails in the top row. Then, keeping your fingers firmly on the original two coins, push them down, so that the two columns of coins move down far enough to make a square again.

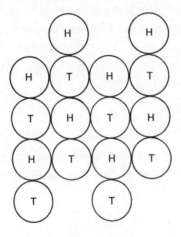

43. Solution:

If the sons divided the land like this, they would each receive an equal-sized L-shaped piece of land.

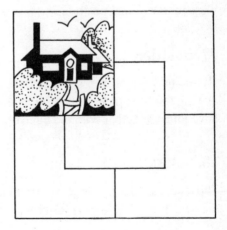

44. Solution:

Mary should have called 'four', making the running total up to 11. Whatever Peter called out on his turn, Mary should always have called enough to make the total up to 11. In this way, the total after Mary's turn would always have been a multiple of 11.

After they had each called nine numbers, the total would be 99, and Peter would be forced to reach 100 with his next call, even if he called 'one'.

45. Solution:

The assistant is selling individual house numbers. Each number costs £1, two of them will cost £2, and so on.

46. Solution:

Tom was planning to pick up the last but one and pour its contents into the second tumbler, then replace the empty glass in its original position.

47. Solution:

He should have broken all three links in one piece of chain, leaving four complete pieces of chain, which he should then have joined together with the three broken links.

48. Solution:

The letters are the first letters of the numbers **one** to **eight**, so the next letter should be **n** for **nine**.

49. Solution:

Each shape has been taken from the same ring, formed by these two quarter circles. You may be surprised that the two pieces have both been taken from the ring on the right. The larger is the top one.

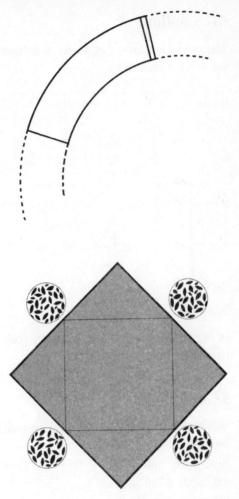

50. Solution:

This is the new pool. The trees remain in place at the middle points of each side, instead of being at the corners.

51. Solution:

The object is a cube, seen from the three directions shown in the second illustration.

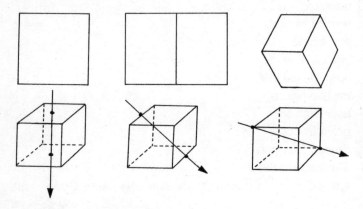

52. Solution:

In order, the answers are: a coffin; your breath; your name; fire; and footsteps.

53. Solution:

Six! To realise why rearranging the pieces will not save any cuts, think of the small cube in the middle of the cube on the table. It has six separate faces, each of which has to be made by a separate saw cut.

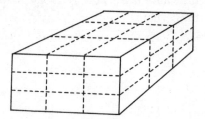

54. Solution:

This is the simplest and prettiest solution:

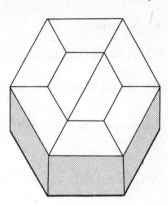

(Mind you, you can probably imagine the arguments about who got the more crumbly middle pieces!)

55. Solution:

The use of a ruler or any straight edge will tell you that the line on the right is actually the continuation of the top line, even though the one on the left might appear to be correct on first impression.

56. Solution:

Four people (including Brian) went to the meeting.

159 is the product of the number of people Brian invited and the number of pages in each document. The only two numbers which, when multiplied together, give 159 are 3 and 53, which are known as its prime factors. So the number of people given documents was either 159, 53 or 3. It is extremely unlikely that Stephen would have had time to distribute 53 documents to 53 people (let alone 159) in the few minutes available, so he must have distributed three copies of the document, and each copy had 53 pages in it. Brian kept the original, making four documents for the four at the meeting.

57. Solution:

FOUR into FIVE: can be done in seven steps, for example:

FOUR POUR POUT PORT PORE FORE FIRE FIVE

Using common words ONE into TWO can be done in ten at most:

ONE ODE ODD ADD AID LID LIP TIP TOP TOO TWO

58. Solution:

The only way to get the two columns to add to the same total is by inverting the number 9, like this:

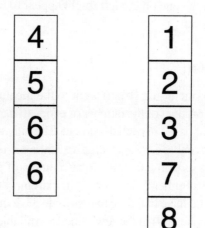

59. Solution:

They don't have lucky friends, since none of the people who won money were their friends! In fact there is very little surprising about this story. Let's take the girl, since she has the report of the biggest win. How many people could her story have come from? There are probably around 1,000 children in the girl's school, each of whom has perhaps ten adult relatives who (if they won the lottery) would count as 'their family'. That makes 10,000 different ways for the story to start. And if in the last four weeks those 10,000 bought 20,000 tickets between them (which is quite possible) then it becomes really quite likely that one family would get a £10,000 win. Stories of big wins spread like wildfire, but nobody is interested in the 19,999 tickets that didn't win! The same applies to the brother's and father's lottery stories.

60. Solution:

The bottle costs £1.05 and the cork costs 5p.

61. Solution:

The trick is very simple: you move onto the long diagonal, from 1 to 2. Your opponent then has to move off the diagonal, allowing you at your second turn to move back onto the diagonal . . . and

END

all the time you are getting nearer to END, which you will be the one to occupy, because it is on the long diagonal.

62. Solution:

They do not move closer together or further apart.

To see why this is so, imagine that there is a block of wood between them. The right-hand bolt is being screwed as if it were screwing into the block, so the wooden block is being pulled towards the right.

The left-hand bolt, however, is being turned as if it were being unscrewed from the same wooden block. In other words, it is releasing it to allow it to move to the right. So the two bolts do not move relative to each other (apart from their separate rotations), and only the imaginary block moves. Take away the imaginary block . . .

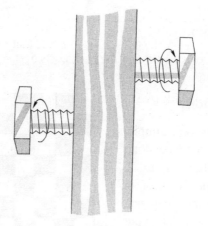

63. Solution:

The reason is simply that the birds spend much longer away from the nest (over 14 minutes) than they do in the nest within sight of Edward Spinks (under 30 seconds), so the chance is more than 29:1 that the first sighting will be a bird flying back to the nest. Since he has only been watching for a week, it is no surprise that so far Mr Spinks' first sightings have always been arrivals. (If Mr Spinks started watching before dawn, of course, he would see a bird leave first.)

64. Solution:

Since all of the statements contradict each other, at least three of them must be false. In fact number 3 is true and the others are false.

65. Solution:

Their conversation included two sentences that include every letter of the alphabet (known as pangrams). These were:

'Pack my box with five dozen liquor jugs', which does it in 32 letters.

'Quick, Baz, get my woven flax jodhpurs', which does it in 30, and is one of the shortest coherent sentences in the English language that uses all 26 letters.

'A quick brown fox jumped over that lazy dog', would contain every letter if it said 'jumps' instead of 'jumped'.

66. Solution:

You cannot float across on top of the planks because then you would certainly be in danger of falling into the water. Instead, arrange the planks at one corner of the moat, as in the diagram.

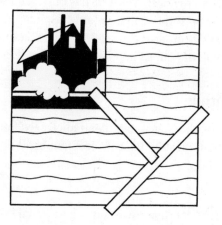

With this arrangement, two planks as short as 3.9 metres can be used to cross a 4-metre moat, allowing for a little overlap where the planks rest on each other or on the bank.

67. Solution:

You cannot write 'one' because then there are two. But if you write two, then there is only one! Lots of words would fit, for example 'small', 'minimal' or 'finite'.

68. Solution:

The difference between 30-all and deuce in tennis is purely psychological: the two scores are exactly equivalent, since to win a game in either situation one player needs to win two points in a row. Whenever a game reaches 30-all the umpire could therefore always call deuce.

Love all, however, is not the same as 30-all. In all serious tennis matches the server is more likely to win a point than the receiver, but it takes a bit of maths (or a lot of experience of playing tennis) to prove that it is more of an advantage to the server (Tuttle) if the score is love all than if it is 30-all. (In fact, if the server's chance of winning any point is always 60%, then his chance of winning the game from love-all is about 74%, while his chance of winning from 30-all falls to 69%.)

69. Solution:

The words fit into the rows and columns and both diagonals like this:

B	A	R	D
A	R	E	A
R	E	A	R
D	A	R	T

70. Solution:

It was 1 January. In any year, Christmas Day is 358 (or 359) days later than New Year's Day, and since neither 358 nor 359 is exactly divisible by seven, the day of the week is always different on these two dates. Damien's birthday is on 31 December, which explains how his birthday can appear to leap by so much. And Janet (who may not be present at this gathering) lives somewhere in the southern hemisphere.

71. Solution:

Thirty-four is the inevitable number of breaks.

Barry starts with one piece of chocolate. Every time he makes a break, he creates one extra piece of chocolate. This is true whether he breaks a small piece into two, or a large rectangle into two parts. Since he ends up with 35 pieces, he must have made 34 breaks.

72. Solution:

To your left. East is to the right when you travel North, but if you are travelling South (which you must be doing if you are leaving the North Pole), East is to the left.

73. Solution:

Potatoes. This has nothing to do with vegetables, though there may be some exotic explanations people try to use about how easy it is to 'scatter' different objects. The explanation is very simple. The words in the sentence have increasing numbers of letters, 1, 2, 3, 4, etc. The gap is a word with 8 letters, of which only potatoes fits from the five supplied.

74. Solution:

The two containers are simply cylinders, so you can half fill either one exactly by tilting it and filling it so that the water just reaches the bottom edge and the top edge of the cylinder, as in this sketch.

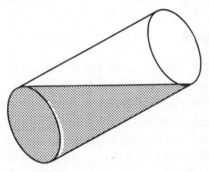

So you can fill each container once, then half fill each once, to transfer a total of 3 litres + 7 litres + 1.5 litres + 3.5 litres = 15 litres in four moves.

75. Solution:

The surprising answer is that the ball would rebound at 120 mph. Its speed relative to the train before impact is 20 + 50 = 70 mph, so the ball will rebound at 70 mph *relative to the train*. But the train itself is travelling at 50 mph, so relative to Craig – and the ground he is standing on – the ball is travelling at a phenomenal 70 + 50 mph. While superballs this bouncy don't exist, this huge amount of acceleration can be observed even by dropping an ordinary bouncy ball onto something moving upwards (like a tennis racket).

76. Solution:

The sequence continues: TH TH TH TH. The letter pairs given in the puzzle are the endings of the ordinal numbers – firST, secoND, thiRD. TH then follows from fourTH to twentieTH.

77. Solution:

There are still 100 olives in each dish, so there must be exactly the same number of black olives in the green dish as there are green olives in the black dish. If this was not true then you would have created olives out of nowhere.

You do not know how many green olives came back in the mixed handful, but you do know there were a total of 20. Let's suppose 17 green and 3 black olives came across.

	Black dish	**Green dish**
Start	100 Black	100 Green
After swap 1	100 B + 20 G	80 G
After swap 2	100 B – 3 B = *97 B*	= *3 B*
	20 G – 17 G = *3 G*	80 G + 17 G = *97 G*

As you see, at the finish in each case the number of green and black olives in their own dishes is 97. In fact, this works for any number in the mixture. Some people still completely disbelieve this answer. The best way to prove it to yourself is to take a real bowl of olives and try it out.

78. Solution:

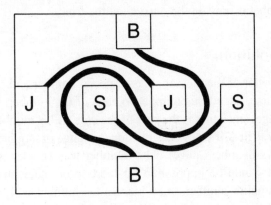

The plots were joined like this. Mr Brown's path was made to wind in and out of the others, and is more than twice as long and twice as expensive.

79. Solution:

One thing James Bond will *not* die of is freezing. All that a fridge does is transfer heat from its inside to its outside, and it uses a lot of electricity to do this. Since the fridge's outside is in the room, the room will in fact slowly warm up. So Bond is at risk of dying of dehydration, suffocation or overheating!

80. Solution:

On the blocks were the letters T W E L V E O N E, which, when rearranged, form the words E L E V E N T W O.

81. Solution:

The label is now on the outside, at the front.

82. Solution:

The writer of the message has a special number and he has, of course, added this number to each of the numbers in the message, since that is the rule of the club. If his number was bigger than 4, it would fall outside the range of numbers that he says members have; if his number was smaller than 4, then it would be impossible for there to be other members with smaller numbers than his. So his number is 4, which means there are 6 members in the club, all with numbers between 2 and 5, two of whom have a smaller number than him, and one with a larger number. All of the numbers add up to 23. The only possible combination of six numbers that fits is 3, 3, 4, 4, 4, 5.

83. Solution:

Arguably there was nothing wrong with her reasoning. This is a version of a well known paradox called 'The unexpected hanging', which has been discussed endlessly by many authors. Here is one way in which you might explain the paradox: it is illogical to tell someone when they will have a surprise. The story could be cut down to Andrew saying: 'Jenny, I'm going to give you a surprise this evening. I will take you for a meal at the Tandoori.' She might logically say 'that's not a surprise because you've told me' so that when he then actually does take her for the meal, she is slightly surprised.

84. Solution:

Here is the simplest way to place the four coins.

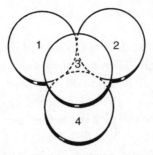

Below is how the five coins are placed. One coin has to be much smaller than the other four. The leaning coins are unlikely to stay in place by themselves – you will have to hold them in place – but that isn't disallowed by the conditions of the puzzle.

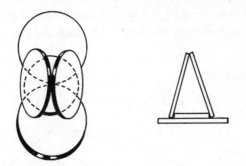

85. Solution:

The cheetah won again! They ran at the same speeds, so that when the leopard had run 90 paces, the cheetah had run 100 and caught up with the leopard. That meant they still had 10 more paces to run to the stream, by which time the cheetah would have nosed ahead.

86. Solution:

The volumes that he does not move will stay in their original order on the shelf. Therefore they must already be in the correct order. So all he has to do is to leave as many volumes as possible which are already in ascending order from left to right. The most he can pick out is four, and he can do this in four ways: 3 4 9 10; 3 5 9 10; 3 4 6 8; 3 5 6 8.

It makes no difference which sequence he chooses. He still has to remove and replace each of the remaining six volumes.

(There is a far harder question, which does not have a simple solution. Since it takes a real effort to push several heavy volumes along the shelf in order to replace the volume he has removed, does the order in which he removes and replaces these six volumes make any difference? The short answer is yes.)

87. Solution:

This is the object, in perspective.

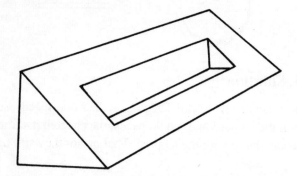

88. Solution:

Yes. The fastest they can do it is seven minutes. Mike puts polish on the shoes of two of the men while Merv puts it on the third man's shoes. Merv then shines Mike's two men while Mike shines Merv's.

Mike: Polish Mr A; Polish Mr B; Shine Mr C (5 minutes)
Merv: Polish Mr C; Shine Mr A; Shine Mr B (7 minutes)

89. Solution:

Old MacDonald's collection of animals consisted of:

PIG (piglet)
KID (goat)
OX (oxen)
SHEEP (sheep)

It is slightly unusual to have an ox on the same farm as a pig, sheep and kid, but that's the way it was. There may also be some more obscure alternatives, such as DEER.

90. Solution:

He turned right instead (naturally, what else could he do?) like this:

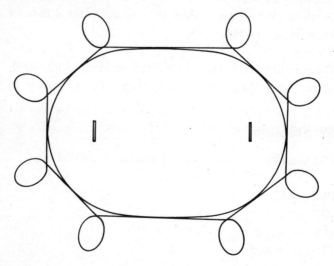

No one asked about how long his curious journey might have been, but the answer would look something like the above. He cycles along the sides of a polygon that can be drawn round the ground, and at each corner, he pirouettes to the right.

91. Solution:

Mrs Smith will overtake all the buses which were occupying positions on her route at the moment she set out, except for those that leave her route during the one hour she is driving along it. She will pass coming in the other direction the same original number of buses plus all the buses which have entered her route during the same one hour. The original number of buses on the route must therefore be 12, with 4 buses leaving and 4 buses joining during her one hour drive, which means that the buses on average leave every 15 minutes.

92. Solution:

The three longest common words that use letters out of QWERTYUIOP are PROPRIETOR, PERPETUITY and (surprisingly) TYPEWRITER. All three of these words appeared in the puzzle.

93. Solution:

Charlie Snod did it.

Suppose the first statement is false. That would mean that 'this statement and the one after the first true statement are both true', which is a contradiction since we said the first was false. So the first statement must be true. This means statement 2 is false, which means statement 3 is true. So (from 2 being false) statement 5 is false. If 5 is false then less than half of the statements are true, and the only way for this to happen is if statements 4 and 6 are also false. So from statement 4, Charlie Snod must be guilty (and Henry van Eyck is not a Dutchman).

94. Solution:

The traditional answer to this is that, since the secretary has established that the third letter is T, there is no reason for him or her to know what it stands for.

However, it is clear from their conversation that they have

a bad telephone line. T for Tummy sounds very like D for Dummy, so it is very natural for the secretary to ask the caller to repeat the T (although a better response would have been 'Sorry, is that T for Tango?').

95. Solution:

Any line through X will divide the bottom row of three buns fairly, and any line through Y will divide the top two fairly, so the line XY is one solution to their problem. Similarly, any line through A bisects the left-hand bun and any line through B (the midpoint of the group) bisects the block of four buns on the right, so the other solution is the line AB.

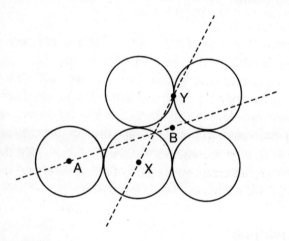

96. Solution:

Gary had heard this trick before and knew that the secret was to put a mat exactly in the middle of the table. Then wherever Marian put a mat, he would put one symmetrically opposite, so that when it became impossible to place a mat it would be Marian who reached this situation first. Unfortunately, in practice it is impossible to place a beer mat exactly at the centre of a large rectangular table without the use of a ruler or a length of string, so if you try this trick, make the bet a small one!

97. Solution:

It can be done in five moves, by inverting (for example) 1 2 3, 4 5 6, 7 8 9, 8 9 10, and finally 8, 9 and 11. Glasses 8 and 9, having each been inverted three times, are finally the correct way up, like the rest.

There are, of course, many other combinations you could use to achieve it in five moves.

98. Solution:

You have to turn over two cards – the lolly card and the vanilla card. If the lolly card doesn't have chocolate on the other side, then Alf is wrong. And if the vanilla card has a lolly on the other side, then he is also wrong. But it doesn't matter what is on the other side of the other two cards: there could be any ice cream on the back of a chocolate card, since what Alf said did not exclude this possibility!

99. Solution:

Charles' numbers were not random. As he wrote down his numbers, he made sure that each digit in his first number when added to each digit of his father's first number always added to 9, and the same for the second numbers.

Dad's first number   7 2 5 8 3 9 1
                               +
Charles' first number   2 7 4 1 6 0 8
                      = 9 9 9 9 9 9 9

That means that the first four numbers would always be equal to 9,999,999 + 9,999,999 = 19,999,998. Then all he had to do to add the five numbers was take dad's third number, add 20,000,000 and take away two – a very simple calculation, which can be done by putting a 2 in front of dad's third number and subtracting two from the end of it.

100. Solution:

The right way round. The easiest way to demonstrate this to yourself is to write on a piece of clear plastic, and look through it at a mirror.

101. Solution:

In order to survive, Sybil needs to run fast enough to remain stationary relative to the wheels. The centre of the wheel is moving forward at 10 mph, but the top of the caterpillar track is moving at 20 mph (while the bottom of the track in contact with the ground is stationary).

So Sybil needs to run at 10 mph towards the back of the tank, in order to be moving forwards at the same speed as the tank. If she can manage more than 10 mph then she will move steadily towards the back of the moving tank and eventually fall off.

102. Solution:

The friend is a woman called Jane. For some reason most people assume that Peter's friend must be a man, even though there is nothing to indicate this in the puzzle, and even though it is very common for college friends to be of the opposite sex.

103. Solution:

This is how it is done. It is not easy to visualise, so we have shown the original cube, and also exploded it to show the three separate (identical) pieces.

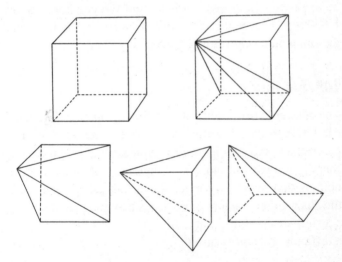

This solution comes from the fact that when you look at a cube along any of its long diagonals, its outline is the shape of a hexagon. The three cuts are across the opposite sides of the hexagon.

104. Solution:

Although he was actually away for only 30 days, he experienced 31 nights: one for each day, plus one for his eastward circumnavigation of the world. When you travel eastwards the sun goes overhead faster so the days and nights get shorter. In Jules Verne's *Around the World in 80 Days*, the plot hinges on the fact that the hero Phileas Fogg experiences 81 'days' on his journey, even though he is away for only 80 days.

What if you were to fly west? Your days would now take longer. Since the circumference of the Earth is about 24,000 miles, if you flew westwards around the equator at about 1,000 mph your nights could last an eternity. Any faster and the sun would start setting in the east!

105. Solution:

The answer is (c) Exactly 50–50. Since you have four coins and I have three, it is certain that you have either tossed *more heads* than I have, or *more tails* (you can't have done both). Since these are the only two options, and since there is nothing that favours a coin to end as tails rather than heads, the chance that you toss more heads than me must be 50–50 – a simple answer to what at first sight looks like it might be a difficult probability question.

106. Solution:

The number opposite 4 is in fact 6. Most people say 3, but this is not an ordinary die because it has two 6s and no 5.

To see why, look at the 2 and 6 in the first two views. In one the 2 slopes to the left, in the other it slopes to the right, so there must be different faces involved. If it is the same 6 in both views then 3 and 4 end up on the same face, so the two 6s must be different. Here is what the whole cube looks like unfolded:

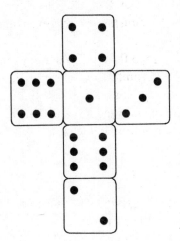